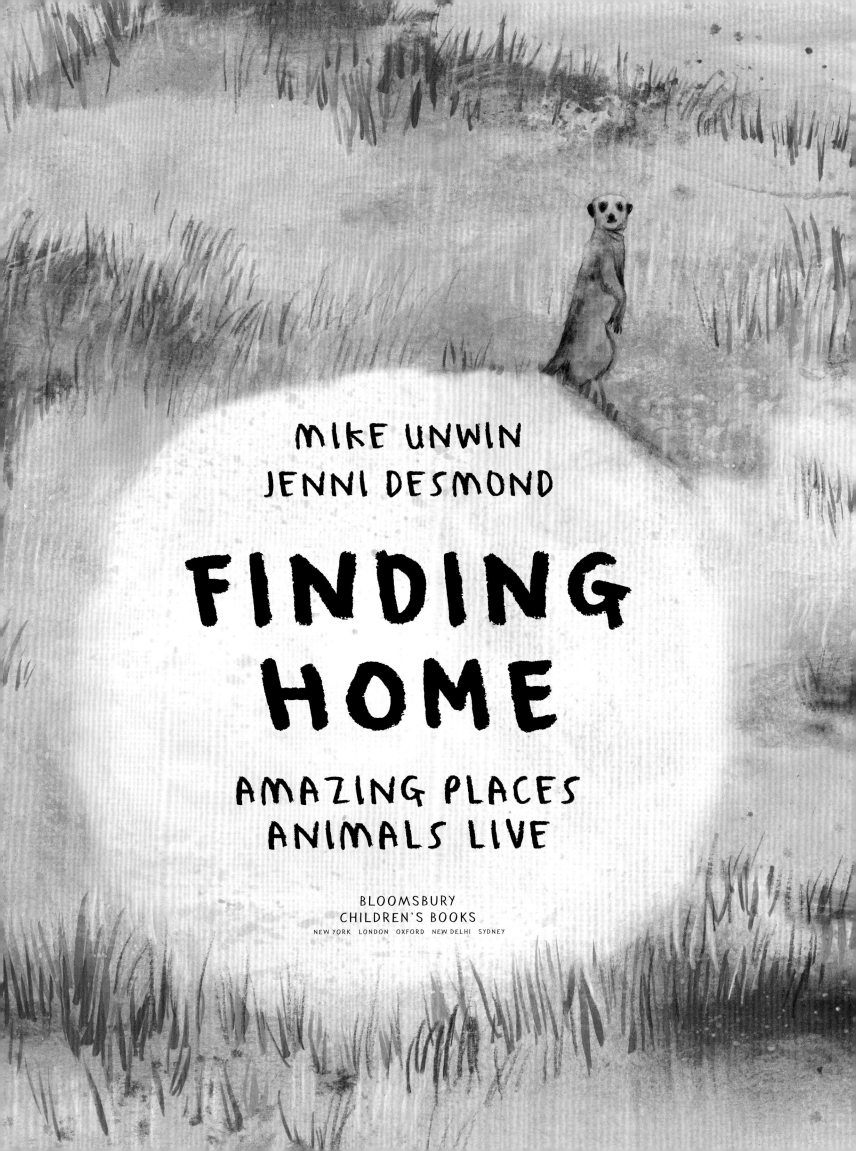

FINDING HOME

MIKE UNWIN
JENNI DESMOND

AMAZING PLACES ANIMALS LIVE

BLOOMSBURY
CHILDREN'S BOOKS
NEW YORK LONDON OXFORD NEW DELHI SYDNEY

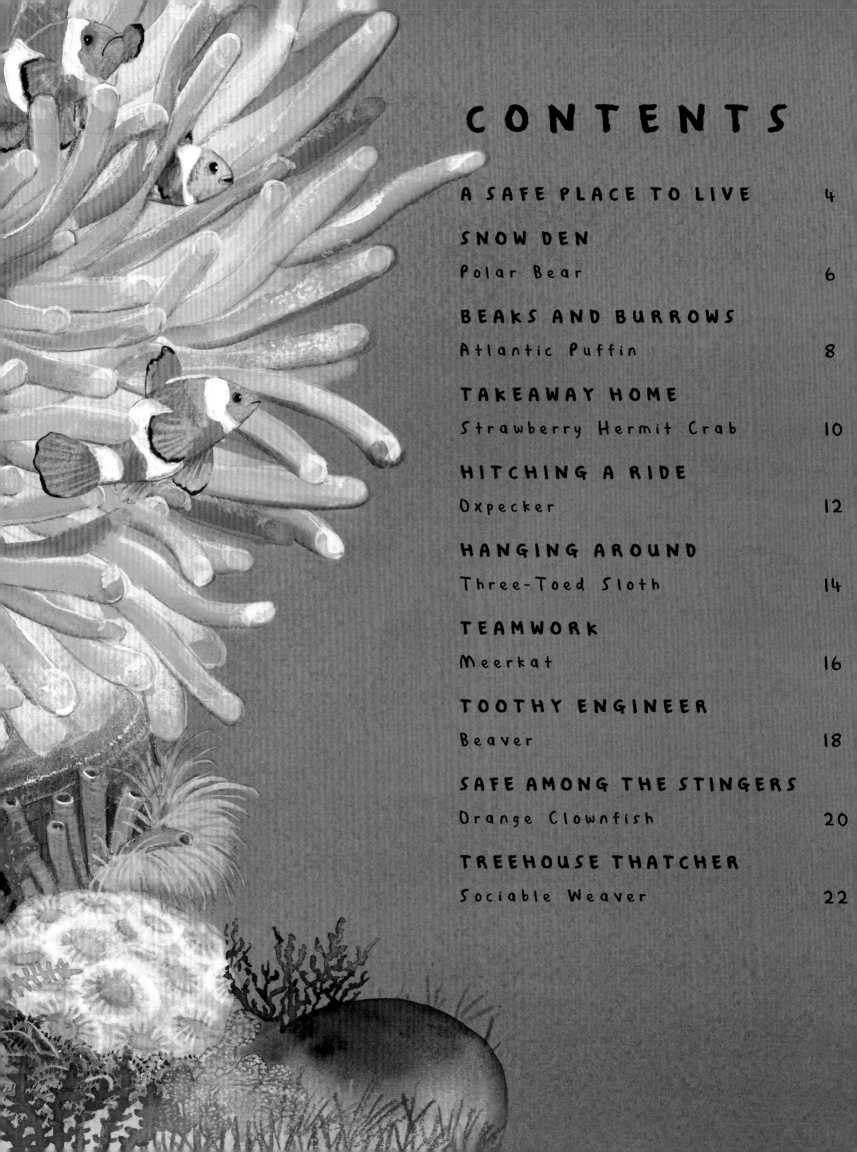

CONTENTS

A SAFE PLACE TO LIVE 4

SNOW DEN
Polar Bear 6

BEAKS AND BURROWS
Atlantic Puffin 8

TAKEAWAY HOME
Strawberry Hermit Crab 10

HITCHING A RIDE
Oxpecker 12

HANGING AROUND
Three-Toed Sloth 14

TEAMWORK
Meerkat 16

TOOTHY ENGINEER
Beaver 18

SAFE AMONG THE STINGERS
Orange Clownfish 20

TREEHOUSE THATCHER
Sociable Weaver 22

HANGING ON
Remora   24

EARTH EXCAVATOR
Aardvark   26

CAVE SWARM
Mexican Free-Tailed Bat   28

CATHEDRALS OF MUD
Cathedral Termite   30

ALL SEWN UP
Common Tailorbird   32

CAMOUFLAGE CHAMPION
Leaf-Tailed Gecko   34

PILING HIGH
Bald Eagle   36

PAPER SKILLS
Common Wasp   38

STUCK INSIDE
Rhinoceros Hornbill   40

MOUNTAIN HIDEAWAY
Snow Leopard   42

JUNGLE TREEHOUSE
Orangutan   44

PLANET HOME   46

A SAFE PLACE TO LIVE

If you could have any home you wanted, what would you choose? Perhaps a treetop castle with a fantastic view, like an eagle's nest. Or maybe a snug underground hideaway, like a meerkat's burrow. Maybe you'd like to share your home with loads of people, like bats in a cave. Or perhaps you'd prefer a place to yourself, like a hermit crab in its shell house for one.

For animals, home can mean many different things. Some live in one place all their lives, like sociable weavers in their huge nest made of grass. Others just stay somewhere for a while, like polar bears in their snow den. Many find a home that has been ready-made by nature, such as a hole in a tree or a cave. Others construct their own, using materials such as grass, sticks, mud, and homemade paper. A few animals even find a home on another animal.

Animals' homes serve many purposes. They can be places to lay eggs, raise babies, hide from danger, keep warm, take a quick nap, or even sleep for the whole winter. But all have one thing common: home is where an animal feels safe. This book explores the homes of 20 different animals around the world. Take a peek inside—you might get a surprise!

Polar bears live north of the Arctic circle. They are the world's biggest land predators. A male can weigh over 1,500 pounds— as heavy as ten adult humans.

SNOW DEN
POLAR BEAR

It's a sunny March morning in the Arctic. Deep under a snowdrift, something seems to be moving. You can hear digging.

Suddenly, out pops a shiny black nose. It's a polar bear! She pushes her head free and takes a deep breath. Then she heaves out her big, furry body and shakes off the snow. She hasn't been outside since October. The fresh air feels good!

Behind her, out pop two more noses—tiny ones—and two pairs of black button eyes. Cubs! These babies are already two months old, but this is the first time they've left the cozy den where they were born.

A female polar bear hunts seals all summer, putting on enough fat to last her through winter. Then, in autumn, she travels inland to find a snowy hillside where she can make her den. She digs a narrow tunnel into the snow and scoops out a chamber big enough to curl up inside.

Winter in the Arctic is dark and cold. But inside her den, the bear is warm and snug. Her thick fur keeps the temperature 77°F warmer than outside. She doesn't need to eat: the fat she built up in the summer helps her survive. Instead, she curls up to save energy. She knows her babies are on the way.

In January, the polar bear gives birth to two tiny cubs, each no bigger than a rabbit. They snuggle up against their warm mom and suckle her rich, nutritious milk, quickly growing bigger and stronger.

Now it's spring and mom is hungry. She hasn't eaten for five months! She needs to return to the coast to catch more seals—and this time she must take her cubs with her. For a few days, she lets them play. Then they all set off together. There's a long journey ahead, and lots to learn.

Beneath all that pale fur, a polar bear's skin is black.

BEAKS AND BURROWS
ATLANTIC PUFFIN

Spring on a puffin island means spring-cleaning for puffins. All winter, the birds have been scattered far and wide across the sea. Now pairs meet at their burrows again—the same ones that they used last year—wagging their heads and rubbing their colorful beaks in greeting.

All puffins nest in burrows. Some dig their own, while others use old burrows dug by other animals, such as rabbits. Either way, after eight months away from home, there's work to be done.

First, each puffin pair chases away any other puffins trying to move in. Then they tidy up, cleaning out any weeds or soil clogging up the entrance. If heavy rain has caused the burrow to collapse, the pair dig a new one. They use their beaks to cut into the grassy hillside, then shovel away loose soil with their orange feet.

Puffins are pigeon-size seabirds that nest on rocky coasts and islands around the north Atlantic Ocean and spend winter at sea. They usually live up to 25 years.

Once the burrow is ready, the female lays her single white egg inside. The two parents take turns looking after the egg, one staying to keep it warm while the other goes fishing. Between fishing trips, they stand guard at the burrow entrance to fend off rival puffins looking for new nests.

Six weeks later, the egg hatches. The chick, called a puffling, stays inside the burrow while its mom or dad goes on trips to catch fish for it to eat. Sometimes, they are away all night. While they're gone, the growing youngster flaps its wings for exercise.

When the chick is five weeks old, the parents head off to start their winter at sea, leaving their chick behind. After a few days, when it realizes no more food is coming, the puffling leaves its burrow. It waits until darkness to avoid hungry gulls, then flaps and tumbles down to the sea below.

Now the puffling must learn to fish for itself. It swims away from the island and won't return for two or three years, when it starts learning how to build a nest of its own. Until then, its home is out on the waves.

Puffins can dive as deep as 200 feet to catch fish under the sea.

TAKEAWAY HOME
STRAWBERRY HERMIT CRAB

A seashell scuttles across a beach. Are your eyes playing tricks? How is it moving so fast? Look closer and you'll find the answer. Underneath, there are tiny legs sticking out. These belong to a strawberry hermit crab. It found the empty shell and climbed inside to make itself a new home.

Unlike other crabs, hermit crabs do not grow their own shell. They are soft and defenseless. But an empty seashell offers great protection, and their curly body is built to slip perfectly into the coiled space inside. They leave their heads and legs sticking out so they can move about and find food, but if they sense danger, they quickly curl back up inside the shell.

It's a clever solution. But there's just one problem: as a hermit crab grows bigger, its shell becomes too small. Soon it needs a larger shell with more space. This can lead to fights, when hermit crabs use their sharp pincers to try to steal each other's shells.

The strawberry hermit crab is just one of more than 800 different hermit crab species worldwide.

To avoid fighting, hermit crabs sometimes come to an arrangement. If one finds an empty shell, it first tries it on for size. If the new shell is too large, it sits and waits. Soon other hermit crabs arrive. They form a line—all waiting for a crab that fits the new shell. When this crab arrives, it clambers out of its old shell and moves into the new one. Now all the other crabs swap shells, each one moving into a new shell that fits. It's a bit like crab musical chairs.

There are many different types of hermit crab. Some live on the seabed; others in rock pools. Some are tiny; others can grow bigger than a coconut. A few even use shells with stinging sea anemones on top for extra protection. But all hermit crabs have one thing in common: they carry their home with them wherever they go.

Some hermit crabs will take their anemones along with them when they move to a new shell.

HITCHING A RIDE
OXPECKER

Imagine a flock of birds clambering up and down your back and pecking at your clothes. Suppose they hopped on to your head and poked their beaks into your hair and ears? And imagine them doing this all day long. How annoying would that be?!

Well, that's exactly what many animals in Africa have to put up with. Antelopes, zebras, giraffes, buffalos, and rhinos often walk around carrying a noisy gang of feathered passengers.

These birds are called oxpeckers. They don't need a ride; they can fly perfectly well. What they want is food. Oxpeckers eat ticks and other tiny bugs that live on animals. Their skillful beaks sift through fur and poke into every wrinkle to find a meal.

One oxpecker may eat up to 300 ticks and 1,000 insect larvae in a single day.

The animals that carry oxpeckers are called hosts. As well as food, a host provides many other home comforts. On its back, the birds take naps, preen their feathers, perform courtship displays, and feed their babies. They even pluck the host's hair for nest material. At night, oxpeckers settle in trees, but during the day their host is their home.

When the breeding season arrives, oxpeckers must leave their host to have their babies. A pair find a tree hole in which to build a nest and lay eggs. As soon as the chicks can fly, they join their parents on the backs of their animal hosts, where they beg for food—lots of juicy ticks and grubs.

Why do animals put up with oxpeckers? It's because the birds help get rid of their irritating skin parasites. If you watch a zebra grazing, you'll see it isn't bothered by the birds—except if they stick their beaks into a painful cut, when it might toss its head or swish its tail. Otherwise, it's happy to provide them with a free home on the move.

HANGING AROUND
THREE-TOED SLOTH

Easy does it. A sloth never hurries. These animals move at such a leisurely pace that they cover, on average, just 118 feet in a day. And for a sloth, that's a busy day! They eat mostly leaves, and leaves can't run away, so what's the point in rushing?

Sloths make their homes in the treetops of tropical rainforests. When they find a good tree with plenty of food, they may stay in it for days or even weeks, eating as much as they need. Around two-thirds of a sloth's life is spent sleeping, slowly digesting its food. During its lifetime, it may visit around 40 different trees.

A sloth's body is specially adapted for its treetop home. Its long claws work like hooks, so it can hang from a branch like a hammock, even when it's asleep. And because it is usually upside down, its fur grows in a downward direction. This helps the rain run quickly off its body—and in the rainforest, it rains a lot!

That's not all that's strange about the sloth's fur. Sloths move so slowly that green algae grows on their hairs, helping camouflage them from treetop predators such as eagles. Moths and beetles even find their own homes in the sloths' straggly coats.

Sloths may be slow but they are brilliant climbers. They move easily through the branches and have no need to come to the ground—except for one reason: to do a poo! This happens about once a week, when they climb down the trunk and do their business at the bottom.

A female sloth gives birth to her baby upside down. The youngster clings to her tummy and rides with her through the branches. After six months, it is ready to find its own home. But it doesn't rush, because what's the hurry?

Sloths live in tropical South and Central America. The different species are divided into two groups, two-toed or three-toed, according to how many claws they have on their front limbs.

TEAMWORK
MEERKAT

Out in the hot African sun, a group of meerkats forages for food. They scratch at the sandy soil with their long claws, hoping to unearth something tasty—a juicy beetle, perhaps, or a crunchy scorpion. Youngsters copy grown-ups . . . until they get sleepy and take a little meerkat-nap.

But not every meerkat is digging. At the edge of the group, one is standing bolt upright on back legs, scanning the horizon. Nearby, another is scanning in the other direction. These are the lookouts. They are looking out for danger.

A sharp alarm bark rings out. All the meerkats look up. One has spotted an eagle. He barks again and dashes toward them. Now the whole gang turns and runs as fast as their little legs will carry them. The big bird swoops down, but—just in time—the meerkats disappear down their burrows. The eagle flaps away, disappointed.

Meerkat burrows have many underground tunnels, with special chambers for sleeping and looking after babies. There are different entrances so the meerkats can get back inside quickly, whichever direction they're coming from—very useful when you're trying to escape an eagle attack.

A group of meerkats is called a mob. They make a great team. Every individual has its job: some work as lookouts; others help look after the babies. Each day, they set out together in search of food and each night, they return to the safety of the burrow. An older female, called the alpha female, is in charge. She decides where the troop will forage every day.

If food runs short—or there is too much danger around—the alpha female may decide to move. The whole colony leaves the burrow and follows her to dig a new one somewhere else. It doesn't take long. That's the beauty of teamwork!

Meerkats belong to the mongoose family. They live in dry areas of southwestern Africa.

Beavers slap their tail loudly on the water to warn their family when danger is approaching.

Beavers are large rodents that live in Europe, Asia, and North America. They eat mainly tree bark and are good swimmers, with thick, waterproof fur. They can live for nearly 20 years.

TOOTHY ENGINEER
BEAVER

Hidden in the forest is a deep, dark pond. In the pond is a big pile of sticks. How did it get there? There are no people around and the water isn't flowing. It's a mystery.

Suddenly, ripples spread across the pond. A furry animal swims across and climbs out on to the bank. It's as big as a small dog but with a face like a guinea pig and a big, flat paddle for a tail. A beaver! Mystery solved! That pile of sticks is its home—called a lodge. A whole beaver family lives there. They built it themselves.

In fact, before the beavers built the lodge, they also built the pond. Once, a fast river flowed through here, but the beavers slowed it down. Using their super-strong teeth, they gnawed through small trees and toppled them to the ground. Then they nibbled the trunk and branches into shorter lengths and dragged them across the river to make a barrier called a dam. By piling on mud and grass, they stopped the river flowing through the dam. It soon spread out behind to form a pond.

Next, the beavers built their lodge in the middle of the pond. They used stones to weigh down sticks underwater, piling on more sticks until they stuck up above the surface. Then they filled in any cracks with smaller branches and mud. Soon, they had a warm, dry home in the middle of the pond.

If you could peek inside a beavers' lodge, you'd see different rooms. There are sleeping quarters, a feeding chamber, and a nursery. In spring, it is in this cozy space that the female will give birth to and raise her babies. Last year's youngsters stay with their parents to help look after the new litter. They also help repair the lodge to keep the whole family safe and snug through the long winter.

SAFE AMONG THE STINGERS
ORANGE CLOWNFISH

Sea anemones may look pretty, but for most small fish, they are death traps. Those colorful, waving tentacles contain deadly poisons. Once they wrap around a fish, there is no escape: soon it disappears into the anemone's mouth, never to return.

So how come the orange clownfish gets so close? This colorful character swims around among the anemone's tentacles without a care in the world. It even hides inside them. That deadly nest is its home.

Don't worry, the clownfish isn't getting hurt! It has a secret seabed superpower. A coating of mucus on its stripy skin protects it from the anemone's venom. This means it can move freely among the tentacles without getting stung. And because it isn't stung, it needn't fear being eaten by the anemone.

On a coral reef, where hungry bigger fish and other predators lie in wait, this superpower is a lifesaver. Among the anemone's tentacles, the clownfish is safe from its enemies, which are scared to get too close. Meanwhile, the clownfish finds an easy meal by polishing off the anemone's leftovers without having to venture too far away.

In return, the anemone gets a free clean-up service. The clownfish nibbles away algae and parasites, and removes any dead tentacles. By swimming around among the tentacles, it helps the water circulate, which keeps the anemone healthy. Some scientists even think that the clownfish's bright colors lure other fish close enough for the anemone to catch.

Clownfish live in small groups around their favorite anemone, which they defend fiercely from other groups. Their bright black, white, and orange markings are easy to spot, so each group can stick close together.

There are 30 different clownfish species. They are 3-7 inches long and live on coral reefs in the Pacific and Indian Oceans.

TREEHOUSE THATCHER
SOCIABLE WEAVER

Deep in the deserts of southwest Africa, a big bundle of straw sits in a tree. Did someone hoist a haystack up into the branches? No! Believe it or not, this is a nest. It belongs to a bird called the sociable weaver.

22

From the huge size of its nest, you'd guess a sociable weaver must be enormous. But this little bird is smaller than a sparrow. Of course, one individual doesn't do the work all by itself. A whole community works together to build one giant nest that houses them all. That's why they're called sociable!

Building takes years. The birds start building their nest by making a platform of strong twigs in the fork of a big tree. They then add stiff, dry grass stalks to build it up. The more they add, the more it clings together. Work never stops: all year round, the birds keep pushing more stalks in. The nest may grow as big as a car, weigh more than a ton, and last 100 years. Some nests become so big that their weight breaks the branch and they collapse to the ground.

Inside, the nest may have more than 100 individual chambers. Each one is home for a different pair of weavers, who come and go through an entrance tunnel underneath. The pair line their chamber with fur, plant fluff, or other soft material to make it cozy. They also stick spiky grass into the tunnel wall to keep out invading snakes. Safe inside, a female lays three or four eggs. Youngsters help their parents feed the new chicks, while neighboring birds without chicks of their own might also come by to feed next door's babies.

Sociable weavers live in their nest all year round. The desert brings blistering hot summer days and freezing cold winter nights, but the dense thatch shields them from these extremes.

Sociable weavers live in the dry, semi-deserts of southwestern Africa. They usually breed after spells of rainfall, when there are more insects to feed their chicks.

Some types of remora have been known to attach themselves to the legs of scuba divers.

HANGING ON
REMORA

Are remoras the bravest fish in the ocean? While most other sea creatures do their best to avoid sharks, remoras get as close as possible. In fact, they literally stick to the deadly predators' sides—which explains why they're also called sucker fish. By attaching themselves to a shark's body, they can travel with it wherever it goes. The shark becomes their home.

It's not only sharks that remoras stick to. These clingy fish also attach themselves to other big sea creatures, including whales, swordfish, manta rays, and turtles. They have good reasons for their daring behavior. First, a free ride saves them valuable energy. Second, they can grab an easy meal by polishing off bits of their host's leftover food, as well as flakes of old skin and even its droppings!

Remoras are found in tropical seas all over the world. There are eight species, and they measure between 12 and 40 inches in length.

But how do remoras get their sticking power? It's thanks to a strange oval disc on top of their heads. This is their sucker. Slats inside it can open or close like a window blind. When open, they create such strong suction that the remora can attach itself to a larger creature using just its head.

A remora may stay attached to its host for up to three months. By sliding backward, it locks on tighter. By sliding forward, it can release itself. Small remoras may even latch on inside the mouths of manta rays, so they can grab food without having to swim about.

Remoras lay their eggs in the open sea, then move on with their host. After hatching, the tiny babies float around among plankton for a year or so, slowly growing bigger. Once they are 1 inch long, their sucker disc is fully formed. They then head out to find a fish to cling to—preferably a big, scary shark.

EARTH EXCAVATOR
AARDVARK

Careful! There's a big hole in the ground. And a heap of soil outside means somebody's been digging. But who? All day long, under the African sun, you watch and wait. But nobody goes in and nobody comes out. This digger must work in secret.

The answer comes after sunset. At first, you hear a shuffling noise inside. Then out peeks a long, twitching nose followed by two big, rabbit-like ears. A strange animal emerges shyly into the moonlight. It has a humped back, a long tail, and big front claws. What on Earth is it?

"Earth" is the right word. This creature is an aardvark—a word from the Afrikaans language of South Africa that means "earth pig." This hole is its home.

Aardvarks feed on tiny ants and termites. By day, they hide away underground. By night, they set out to find food, sniffing out anthills with their sensitive nose and breaking through the hard soil with their claws. When the furious insects swarm out, the aardvarks use their long, sticky tongues to lick them up.

The aardvark is Africa's champion digger. Those powerful claws can excavate an underground burrow up to 43 feet long. Inside, there are sleeping chambers and a special nursery chamber where a female aardvark has her baby. After two weeks tucked up underground, the youngster starts to follow its mom outside. At six months, it can dig a burrow all by itself.

One aardvark can eat an amazing 50,000 ants and termites in a single night!

Aardvarks move home regularly. Whenever they move out, other animals move in. Abandoned burrows provide cozy homes for all sorts, from pythons and porcupines to wild dogs and warthogs. Some aardvark holes are so big that you could easily crawl inside. But be careful: you never know who you might meet!

Aardvarks live in open grasslands across much of Africa—wherever they can find enough insects to eat. They can weigh over 130 pounds and live for over 20 years.

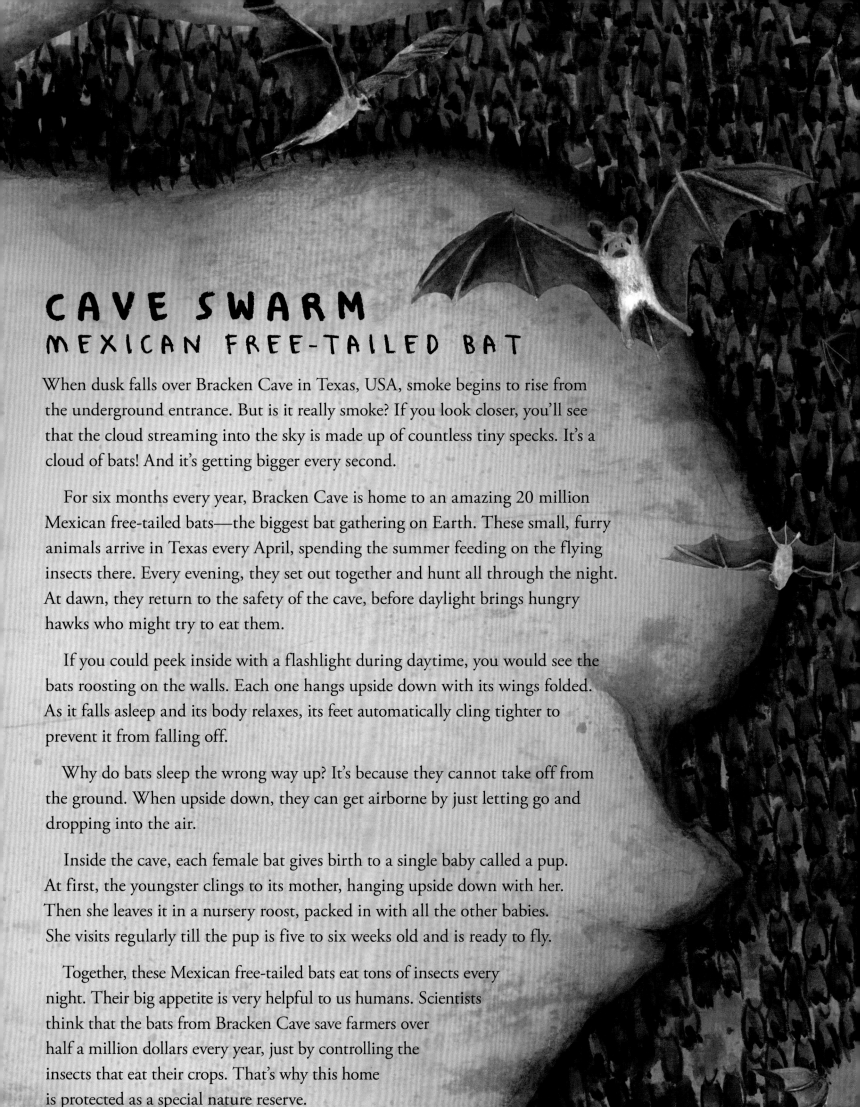

CAVE SWARM
MEXICAN FREE-TAILED BAT

When dusk falls over Bracken Cave in Texas, USA, smoke begins to rise from the underground entrance. But is it really smoke? If you look closer, you'll see that the cloud streaming into the sky is made up of countless tiny specks. It's a cloud of bats! And it's getting bigger every second.

For six months every year, Bracken Cave is home to an amazing 20 million Mexican free-tailed bats—the biggest bat gathering on Earth. These small, furry animals arrive in Texas every April, spending the summer feeding on the flying insects there. Every evening, they set out together and hunt all through the night. At dawn, they return to the safety of the cave, before daylight brings hungry hawks who might try to eat them.

If you could peek inside with a flashlight during daytime, you would see the bats roosting on the walls. Each one hangs upside down with its wings folded. As it falls asleep and its body relaxes, its feet automatically cling tighter to prevent it from falling off.

Why do bats sleep the wrong way up? It's because they cannot take off from the ground. When upside down, they can get airborne by just letting go and dropping into the air.

Inside the cave, each female bat gives birth to a single baby called a pup. At first, the youngster clings to its mother, hanging upside down with her. Then she leaves it in a nursery roost, packed in with all the other babies. She visits regularly till the pup is five to six weeks old and is ready to fly.

Together, these Mexican free-tailed bats eat tons of insects every night. Their big appetite is very helpful to us humans. Scientists think that the bats from Bracken Cave save farmers over half a million dollars every year, just by controlling the insects that eat their crops. That's why this home is protected as a special nature reserve.

Cathedral termites live only in northern Australia, though other mound-building termite species can be found in Africa and South America.

CATHEDRALS OF MUD
CATHEDRAL TERMITE

If you travel across northern Australia, you might be surprised to see strange, knobbly towers looming above the grass. From a distance, they look like miniature cathedrals. Some are 20 feet tall. But who built them?

Believe it not, these towers were built by tiny insects called termites, each no bigger than a grain of rice. Termites live in huge colonies, often millions strong. They build their towers over many years, working closely together. One worker deposits a tiny blob of damp soil. The next adds another blob to it—and so on. Gradually, the structure begins to rise. Eventually, the termites have built a huge tower, baked hard as concrete by the tropical sun.

It might surprise you to learn that the termites don't actually live inside the mound. Their nest is underneath. Here, in a maze of passages and chambers, the termites live and work together. Only the queen has her own chamber. Inside, she lays up to 1,000 eggs a day. Meanwhile, workers feed the babies with a special fungus that they grow on chewed-up wood pulp, and soldiers dash about protecting the nest.

The nest can get very stuffy. This is where the tower does its work. Inside, the termites have built lots of tunnels, like pipes. When the sun heats the outer walls, the warm air inside rises up the pipes near the surface and then falls down a tunnel at the center—before rising back up again. This keeps air circulating. It creates perfect conditions for the termites: not too hot, too cold, too wet, or too dry.

So, just like humans, termites build skyscrapers. But unlike us, they don't need to burn carbon to keep them warm and ventilated. Their amazing architecture uses the natural energy of the sun. Scientists think these insect engineers can teach us a lot about green energy.

Although termites live in organized colonies like ants, they are more closely related to cockroaches.

Tailorbirds belong to the warbler family. The common tailorbird is the best known of 16 species. All live in tropical Asia, from India to Indonesia.

ALL SEWN UP
COMMON TAILORBIRD

Did you think that it was people who invented sewing? Well, look at a tailorbird's nest and think again! This little bird gets its name for a good reason. Working just like a tailor, it stitches together a perfect nest for its eggs and chicks.

Tailorbirds don't have fabric to work with, of course. Instead, they use leaves. Each pair works together. First, they find a suitable nest site, deep in a bush and hidden from predators. Then, the male collects fine material such as plant fibers and spider silk. These threads will form the stitching that holds the nest together.

Next, the female gets to work. She uses her sharp little beak to punch tiny holes around the edge of each leaf. Finally, she draws the plant fiber and spider silk through the holes to pull the edges together—just like a tailor using needle and thread. Now she has a perfect cradle. Sometimes, she may sew up one large leaf to do the same thing.

The cradle is just a bit bigger than your hand, and sewn together with the upper side of the leaf facing outward. This makes it harder for predators to spot. Inside, the birds build a snug nest using soft materials collected from other plants. Here, the female lays three to five eggs, which she incubates for 12 days. Only three weeks after hatching, the chicks leave the nest. They often roost together with their parents, sandwiched tightly between them on a thin twig.

The tailorbird often finds a home in parks and gardens, where it forages around borders and flowerbeds for insects. With its cheerful little song and tail held high, this is an easy bird to spot. But try as you might, you'll never find its clever little nest.

Some cuckoos lay their eggs in tailorbird nests, tricking the tailorbirds into raising the cuckoo chick instead of their own babies.

Geckos don't have eyelids—they clean their large eyes with their long tongues.

CAMOUFLAGE CHAMPION
LEAF-TAILED GECKO

There's a lizard on that tree trunk. Can't you see it? Look closer. Study the cracked bark and peeling lichen. Can you make out tiny toes and a big flat tail? And those two round lumps: aren't they a pair of big, staring eyes?

The leaf-tailed gecko makes its home deep in the rainforest of Madagascar—a large island in the Indian Ocean. All day long, it lies flat on a branch, head pointing downward, not moving an inch. Birds, snakes, and other enemies have no idea it's there, even though it's right in front of their eyes.

How can anything hide in plain sight? Easy! The gecko's body is the exact color and pattern of the tree it rests on, right down to the moss and lichen. It flattens itself so nothing sticks out and no shadows give it away. Flaps of skin around its body help blend its outline into the tree. In a camouflage Olympics, this lizard would win gold!

Different types of leaf-tailed gecko are adapted to different backgrounds, depending upon where they live. Some look like branches, others like leaves. Many can change color to match their background as they move from place to place. Once they find the perfect spot, it becomes their home for a while and they return to it day after day.

If a predator such as a bird sees through the disguise, the gecko has one more trick. It lunges forward, gaping wide its bright orange mouth and hissing like a snake. The bird gets a shock. Before it can recover, the gecko dashes away to safety.

Otherwise, a leaf-tailed gecko doesn't move until nighttime. Then, it sets out through the tangled greenery in search of its favorite foods: beetles, crickets, and other crunchy creepy-crawlies. The poor bugs never see it coming—until it's too late!

Leaf-tailed geckos live only on Madagascar. There are 21 different species. They range in length from 4 to 12 inches.

PILING HIGH
BALD EAGLE

How high can you build a Jenga tower? You'd do well to top a bald eagle's. This big American bird of prey stacks up branches in a tree to make the biggest stick nest in the world. One record-breaking eagle nest in Florida was over twice the height of a man and weighed two tons—as much as a small car.

Bald eagles eat mostly fish, so they usually nest near water. A pair find a tall tree and work together to build their nest—called an aerie—near the top. They start by gathering big sticks, which they carry and arrange carefully on a platform of strong branches. As they add more and more sticks, the weight holds the structure in place.

The bald eagle is the national bird of the USA.

The bald eagle is not really bald but has a head covered in white feathers. Its wingspan may measure up to 7 feet.

The female weaves in smaller sticks and twigs to fill any gaps. At the center, she creates a small bowl, which she lines with moss, grass, and even seaweed. The nest needs to be comfy inside: she will soon be sitting on her eggs.

The chicks—usually two—hatch after about five weeks and spend 10 to 12 weeks in the nest. At first, one parent stays with them while the other goes off to find food. As the chicks grow bigger and stronger, both parents make fishing trips. While the adults are away, the youngsters flap their wings as they build up strength for their first flight.

A pair of bald eagles may use the same aerie for many years. They prepare for each new breeding season by cleaning it out and repairing any holes. By adding new sticks, they make the nest even bigger than the year before.

Sometimes, winter storms blow the nest down—or it grows so big that the branches break. If this happens, some eagle pairs will move to a "backup nest" they built for this very situation. Otherwise, they'll just stay where they are and put their old home back together again, piece by piece.

PAPER SKILLS
COMMON WASP

Have you ever tried origami? Making things with paper is tricky. But wasps are experts. Not only do they construct beautiful paper nests: they even make their own paper!

It is the queen wasp that starts to build the nest. Her work begins in spring, when she wakes up from her hibernation. First, she finds a sheltered site, such as a hole in a tree, a rock crevice, or even inside an attic. Then, she chews up plant and wood fibers to make a squishy pulp. She plasters this on to the roof of the hole to form the anchor of her nest.

Now, she builds downward, adding more pulp until she has made a stalk—called a petiole. On the end of the petiole, she builds several six-sided chambers, called cells. Soon the pulp dries out into a thin papery substance. Then she lays an egg in each cell.

When the first eggs hatch, the queen feeds the larvae with caterpillars and other insects. These first turn into cocoons then hatch into worker wasps. The queen keeps laying eggs while the workers build more cells, making the nest bigger. They also take over the job of feeding the larvae.

The nest keeps growing. Eventually, it may reach the size of a soccer ball and become home to thousands of wasps. The workers look after the growing youngsters. They use their fierce stings to protect the nest and smear a special chemical around the petiole to stop ants from getting in.

Wasps may not be our favorite animals—especially when they buzz around our picnics—but these amazing insects do an important job. The adults feed on nectar, so when they fly from flower to flower gathering this sticky, sweet food, they help pollinate our plants. This makes wasps an essential part of our ecosystem.

There may be more than 100,000 wasp species in the world. The common wasp is found across Britain, Europe, and Asia.

STUCK INSIDE
RHINOCEROS HORNBILL

For a female rhinoceros hornbill, home must feel like prison. Every summer, she is trapped in a tree trunk for four whole months. Inside, it's so cramped that she can barely turn around. The only daylight comes from a tiny slit, just big enough to poke out the tip of her huge beak.

You might wonder why any bird would make itself so uncomfortable. For the hornbill, it's about raising babies. This big bird lives in tropical rainforests, where eagles, snakes, and other predators lurk in the greenery. She and her mate use this special way of nesting in order to keep their chicks safe.

The pair begins work at the start of the breeding season. First, they find a suitable hole in an old tree trunk. Then they use their big, strong beaks to make it bigger. Once they have chipped out a cavity large enough, the female climbs inside and settles down to lay her two eggs.

At first, the female can hop out from time to time for fresh air. But once the eggs have hatched, the male brings beakfuls of mud and the pair plaster up the entrance. Soon, the female is trapped inside the hole by a wall of dried mud. The male can pass her food through the slit, but there's no way predators can get in.

Baby rhinoceros hornbills take 11 to 12 weeks to grow up. They start out featherless and helpless, but as the male brings more food—mostly fruit and insects—their feathers sprout and they quickly grow bigger and stronger.

By the time the chicks are ready to fly, life inside the hole has become very squashed. Now the female bashes down the mud wall with her beak, squeezes through the gap, and joins her mate outside. Freedom! Together, they encourage their babies to head out into the big wide world.

The rhinoceros hornbill lives in Southeast Asia. It measures about 35 inches—as long as a baseball bat—and can live for 35 years.

The rhinoceros hornbill gets its name from the lump on its beak, called a casque, which looks like a rhino's horn.

MOUNTAIN HIDEAWAY
SNOW LEOPARD

It's no secret that cats are good at climbing. But one big cat takes it to extremes. The snow leopard's home is in the Himalayas, the world's highest mountains. Here, it wanders the peaks and canyons, reaching heights of up to 20,000 feet: that's 20 times the height of the Eiffel Tower!

Of course, the snow leopard is not just any cat. This furry mountaineer is built for life at high altitudes. Its broad paws prevent it from sinking into snow and its long tail helps it balance when leaping over cliffs and boulders.

A snow leopard's thick fur coat keeps it warm and cozy at freezing temperatures. When it curls up to sleep—finding a snug, hidden place among the rocks—its long, furry tail wraps around its face like a woolen scarf.

Snow leopards can leap over 30 feet in a single bound.

The coat also provides amazing camouflage. When the cat sneaks up on its prey, such as mountain goats, its creamy gray color and smudgy spots disappear against the background rocks and snow. Prey may stare right at a snow leopard without seeing it.

Snow leopards have huge hunting territories: one individual may cover over 600 square miles, moving regularly from place to place. But in spring, a female must find a safe place to have her cubs. A cave or crevice makes the perfect mountain den. She chooses one with a hidden entrance and a ledge outside, from where she can look out for prey—and danger. Inside, she lines the floor with soft fur from her tummy to make it snug for her babies.

A female usually has two or three cubs. At birth, they are blind and helpless. One week later, their eyes open, and by week five, they can walk. When their mother goes away on hunting trips, she knows they will be safe in their hideaway until she returns.

Young snow leopards stay with their mother for 18 months, learning how to survive. When it's time to leave, they climb high into the mountains to find a home of their own.

Snow leopards are a different species from other leopards, with shorter legs and a longer tail. They live only in the Himalayas and other mountains in Central Asia.

Orangutans live up to 45 years, and are native to Borneo and Sumatra in Southeast Asia.

JUNGLE TREEHOUSE
ORANGUTAN

Imagine: it's dawn, and you're tramping through the jungle. Suddenly, you spot a big nest in a tree. What kind of huge bird made this, you wonder? Then a long, red hairy arm reaches out and a puzzled face peers at you through the leaves. That's no bird! You've found the nest of an orangutan, the tree-climbing ape of Asia.

You were just in time. The hairy primate grabs a branch and swings out of her nest. Surprise! There's a sleepy baby clinging to her tummy. The two apes disappear into the greenery. They are constantly on the move through the forest in search of wild fruit—and by scattering seeds as they go, they work like gardeners, helping trees and other plants to grow.

At the end of each day, an orangutan must find somewhere to spend the night. The ground isn't safe, so it builds a nest high up in the trees. Once it's found a suitable spot, it bends down some big branches and locks them together into a platform. Then it arranges small branches to make a mattress on top, weaving in loose ends with its clever fingers. Finally, it bunches up some leafy branches for a pillow and pulls more leaves over for a blanket. Perfect! A snug bed for the night.

When a mother orangutan is busy building a nest, her baby watches and learns. By the age of three, it will be able to make its own nests—though it will still stay with mom for up to eight years.

Orangutans make a new nest every night. But as people cut down more and more trees in their forests, it is becoming harder for them to find safe places to sleep. We need to protect the places where orangutans live.

After all, these gardeners have important work to do.

PLANET HOME

Our planet is home to a mind-boggling number of different animals: at least 6,500 species of mammal, 10,000 species of reptile, 11,000 species of bird, and literally millions of insects. Every single one has its own home. And each home forms part of a natural habitat, such as a forest, desert, or ocean.

All these habitats, and the animal homes that they protect, fit together like jigsaw pieces to make up planet Earth. And just like a jigsaw needs all its pieces, a healthy planet needs all its animals and their habitats. This is important for us, too.

In Canada, beavers help prevent flooding. Their dams slow down rivers so that water spreads out and sinks before it floods towns.

In Texas, Mexican free-tailed bats eat billions of insects nightly, stopping them from damaging crops.

Sadly, we human beings have not been looking after our planet. We have cut down forests, drained wetlands, and dumped waste in the oceans. And by burning fossil fuels such as coal and oil to power cities and factories, we have raised our planet's temperature, killing coral reefs and melting glaciers and sea ice. Through our careless behavior, countless animal homes are disappearing and many species are in danger.

Today, scientists and conservationists are working hard to put things right, by protecting wildlife, restoring habitats, and finding more sustainable ways of using the planet's resources—for example, through recycling plastics and using renewable energy.

There is only one planet Earth. To keep it safe and healthy for ourselves, we need to look after it for all other animals, too. We must protect the habitats where animals live and not take more than we need. Only then will planet Earth truly be home, sweet home.

In Europe and all over the world, wasps do vital work pollinating plants and reducing the number of insect pests.

In the mountains of the Himalayas, snow leopards keep wild sheep and goat numbers stable, stopping them from overgrazing native plants and damaging their ecosystem.

In Borneo, orangutans are guardians of the rainforest. They spread seeds, make the soil fertile, and help plants to grow.

In Africa, oxpeckers help protect rare rhinos from poachers by sounding their alarm call whenever humans approach.

In Australia, cathedral termite mounds have taught us to make better use of the sun's energy for heating and air-conditioning.

For my daughter Flo, in all the homes she ever finds —M. U.

For Daffles with love —J. D.

BLOOMSBURY CHILDREN'S BOOKS
Bloomsbury Publishing Inc., part of Bloomsbury Publishing Plc
1385 Broadway, New York, NY 10018

BLOOMSBURY, BLOOMSBURY CHILDREN'S BOOKS, and the Diana logo
are trademarks of Bloomsbury Publishing Plc

First published in Great Britain in June 2024 by Bloomsbury Publishing Plc
Published in the United States of America in March 2025
by Bloomsbury Children's Books

Text copyright © 2024 by Mike Unwin
Illustrations copyright © 2024 by Jenni Desmond

All rights reserved. No part of this publication may be reproduced or transmitted in any form
or by any means, electronic or mechanical, including photocopying, recording, or any
information storage or retrieval system, without prior permission in writing from the publisher.

Bloomsbury books may be purchased for business or promotional use. For information on bulk purchases please contact
Macmillan Corporate and Premium Sales Department at specialmarkets@macmillan.com

Library of Congress Cataloging-in-Publication Data
available upon request
ISBN 978-1-5476-1510-0 (hardcover) • ISBN 978-1-5476-1511-7 (e-book) • ISBN 978-1-5476-1512-4 (e-PDF)

Art created with watercolor, acrylic, ink, pencil, and pencil crayon
Typeset in Adobe Garamond Pro and JenniDesmond
Book design by Katie Knutton
Printed in China by C&C Offset Printing Co. Ltd., Shenzhen, Guangdong
2 4 6 8 10 9 7 5 3 1

To find out more about our authors and books visit www.bloomsbury.com and sign up for our newsletters.